BEGINNING ALGEBRA THINKING
For Grades 5-6

JUDY GOODNOW

No. 7271

Beginning Algebra Thinking, Grades 5 - 6

Limited Reproduction Permission: Permission to duplicate these materials is limited to the teacher for whom they are purchased. Reproduction for an entire school or school district is unlawful and strictly prohibited.

Judy Goodnow is an author, curriculum developer, and editor for Ideal School Supply Company. She has taught children at the kindergarten level through grade six, and was a language arts and math resource teacher.

As a curriculum developer, Judy has authored or coauthored over 100 books and sets of games and activities for reading, language arts, and math. In addition, she has developed several language arts and mathematics software programs. She has conducted workshops for teachers in the use of math manipulatives and computer-related materials.

Judy holds a bachelor of arts degree from Wellesley College, a master's degree in Interactive Educational Technology from Stanford University, and earned her California Teaching Credential at San Jose State University.

Illustrations by Donna Reynolds • Graphic Design by Annelise Palouda • Project Manager: Nancy Tseng

© 1994 Ideal School Supply Company • Oak Lawn, Illinois 60453 • Printed in U.S.A.

ISBN: 1-56451-096-4

3 4 5 6 7 8 9 10. 9 7 6 5 4

Table of Contents

Notes to the Teacher	iii
Solutions	vii
Number Combinations	1
Number Puzzles	7
Seesaw Balances	13
Ratios and Number Combinations	19
Number Patterns	25
Fraction Puzzles	31
Linking Circles	37
What's the Rule?	43
Story Problems	49

Notes to the Teacher

This is one of two books designed to prepare students for thinking algebraically:

Beginning Algebra Thinking, Grades 3-4
Beginning Algebra Thinking, Grades 5-6

The National Council of Teachers of Mathematics (NCTM) stresses that by providing students with "informal algebraic experiences throughout the K-8 curriculum, they will develop confidence in using algebra to represent and solve problems." Many algebra students never understand that algebra is a problem-solving tool, because for them, algebra is merely a group of symbols to be manipulated. They don't see the relevance of algebra to their world. If, in the early grades, we can give students rich experiences with problem-solving strategies, and introduce them informally to equations, variables, and other algebraic concepts, they will build the foundation for understanding this powerful tool.

Each book presents 54 reproducible pages of problems—some designed to be solved using concrete manipulatives (cubes), some designed to be solved using a calculator, and some presented in story form. Solutions are provided, including sample algebraic expressions and equations.

Introduction to **Beginning Algebra Thinking, Grades 5-6**

This book provides an informal introduction to thinking algebraically, beginning with problems in which students use cubes to find solutions. Manipulating concrete objects gives students the opportunity to explore problems, trying out different solutions. Exploring problems in this way allows students to develop a visual image of the solution process, making it easier to solve the same problem in a more abstract context. Next, students solve the same type of problem using a calculator. The calculator is an important tool for students, allowing them to concentrate on the problem-solving process. Finally, the same type of problem is presented in a story problem format. It is helpful for students to solve the same type of problem in a variety of formats.

Students solve each type of problem using a problem-solving strategy or combination of strategies. These strategies include: acting out with objects, working backwards, making and using tables, making and using diagrams, making an organized list, guess and check, and using patterns.

These strategies help students learn to identify the unknown quantity (variable) of a problem, and to see the relationship of the unknown quantity to the other data in the problem. These thinking skills are crucial for translating word phrases into algebraic expressions and equations—essential steps in algebra. As machines become more efficient with symbol manipulation, the translation from words to symbols becomes a critical skill for our students to master. All of the activities in this book allow students to practice this translation process.

Using **Beginning Algebra Thinking, Grades 5-6**

Contents

This book contains eight sequences of problems. Each six-page sequence focuses on one type of problem. The first four pages present problems accompanied by a diagram on which students can move cubes to work out solutions. The fifth page presents problems of the same kind containing larger numbers; and students use a calculator to solve them. The sixth page presents problems of the same kind in story form.

After the eight sections, there are six pages of mixed story problems. Problem types are mixed so the student will have to identify the strategies needed to solve each problem.

Suggestions for Classroom Use

The problems are sequenced according to level of difficulty within each section. If you find that a problem or section is too challenging for your students, or not challenging enough, you can modify it to meet their needs.

These problems can be used by students working in pairs or individually. Working together encourages students to talk about their thinking and their discoveries. It is beneficial for students to articulate their thinking and to hear how others may have solved the same problem in a different way. Encourage your students to share their ideas with other pairs of students, with other small groups, or with the whole class. You may want pairs of students to show how they solved a problem by using the overhead projector. This talking about the process helps students make mathematical connections and enriches their understanding.

You can use the section of mixed story problems in a variety of ways. After students have completed the problems in a section, you can have them do the additional problems on pages 49-54 that are the same type. You may want to wait until students have completed all the sections. Then they can think about which problem-solving strategies are appropriate for each problem. Have students create their own story problems. Pairs can write problems and exchange them with another pair, without identifying the type of problem.

Materials

It is recommended that each student or pair of students have a pencil, 100 cubes (20 of five different colors) for the pages showing the cube icon, and a calculator for the pages showing the calculator icon. You can make copies of the problems for each student or pair of students.

Introducing the Problems

Go over the first problem in each section with the students, before they begin work. Review the problem-solving strategies used in that section. Help the students identify what information they are looking for in the problem. Have them identify what information is known and what information is unknown.

You might want to make an overhead transparency of the page and present it on the overhead projector using transparent colored tiles. After the students have worked out some of the problems, you may want to have them demonstrate their solution strategies on the overhead.

Encourage your students to try writing algebraic expressions and equations for the problems. First, give them time to explore with the cubes, then talk about how they could translate the information or data in the problem into letters or symbols. Talk about a *variable: a letter, shape, or other symbol used to represent an unknown quantity*. Also talk about an *equation: a number sentence in which the expressions on both sides of the equal sign represent the same value*. The solution section shows ways in which students can write algebraic expressions and equations for the problems. Students may find other ways to write expressions and equations that are equally correct.

Following is a brief description of the types of problems in each of the eight sections and the problem-solving strategies students can use to solve them.

Pages 1-6: Number Combinations

Strategies students can use: act out with objects, guess and check, make an organized list, use a diagram, and use a calculator.

Students are given the total number of objects, the number of different colors, and clues about the number relationships between the colors. The variables are the unknown number of each color. Students will begin by guessing and checking. The students take the cubes and explore representing a color with different numbers of cubes. They look for all the ways they can solve the problem. As they make a list of their answers, they will begin to see relationships between the numbers.

The students can write the clues as algebraic expressions. For example, for the first problem on page 1, they can write: $R > Y$. They could also write an equation for this problem: $R + Y = 8$.

Pages 7-12: Number Puzzles

Strategies students can use: work backwards, act out with objects, use a diagram, and use a calculator.

Students begin with one known quantity, which is given at the end of the problem. They work backwards, using the clues for each unknown to find its value. They can use different-colored cubes to represent each thing, beginning with the known quantity.

The students can write equations for each clue. For example, equations for the clues in the first problem on page 7 could be: $C = 4$, $N = C + 2$, $P = N + 3$. An equation for the total could be: $C + N + P =$ Total.

Pages 13-18: Seesaw Balances

Strategies students can use: act out with objects, use a diagram, guess and check, make and use an organized list, and use a calculator.

A balance is a good model for introducing students to equations, because both sides must have the same mass (equal value) for it to balance. In these problems, each robot on the seesaw has a weight (represents a number). Robots that are the same shape have the same weight. Robots that are different shapes have different weights. The total weight of robots on one side must equal the total weight of the robots on the other side. Students are given the total weight of all the robots on the balance. They can take cubes equal to the total weight and find out how many ways they can arrange the cubes so that the sides are equal. Recording in the list will help students to see whether all possible solutions have been found. They can also look for patterns in the relationships between the numbers in the list.

Students can write equations for these problems. For example, an equation for the first problem on page 13 could be: $3 \hexagon + \bigcirc + \triangle = 12$. They could also write an equation to show the relationships between the shapes: $\hexagon + \bigcirc = \triangle$

This may be easier to see in problem 2 where there are three possible solutions.

Pages 19-24: Ratios and Number Combinations

Strategies students can use: act out with objects, make and use a table, use a diagram, guess and check, and use a calculator.

Students begin by making a table of possible number combinations based on the ratios between numbers of antennas and given creatures. Then they can guess and check, using the cubes and different number combinations from the table to find the combinations that sum to a given number. They can look for patterns that will help them be sure they have found all the solutions.

Students can write an equation for these problems. For example, an equation for the first problem on page 19 could be: $(1 \times \textit{Zim}) + (2 \times \textit{Ziff}) = 12$.

Pages 25-30: Number Patterns

Strategies students can use: act out with objects, use a diagram, make and use a table, look for patterns, and use a calculator.

Students use cubes to represent the number of plants and creatures that Biologist Zingbat and Zoologist Whizzit discover. They record the numbers in a table. They look for a pattern in the numbers of cubes and in the numbers recorded in the table. The pattern may be a pattern of increase or a pattern of decrease. They use the numerical pattern to predict how many creatures or plants they will find on the fifth or sixth day.

Students can generalize a rule about the numerical pattern and then use the rule to predict the number on the fifth or sixth day. In the first problem on page 25, for example, the number of fizzywits increases by two each day and the number of moosles increases by four each day.

Students can write equations for the rules. For example, for the first problem on page 25 where the fizzywits increase by two and the moosles by four, the equations could be $F = 2n$ for the fizzywits and $M = 4n + 1$ for the moosles, and n stands for the number of the day.

Pages 31-36: Fraction Puzzles

Strategies students can use: make and use a diagram, act out with objects, and use a calculator.

Each problem gives information about a fractional part of the total. Using the diagram will help students find the total number and then find the value of each variable.

Students can write equations that show how to find the total and the variable. For example, for the first problem on page 31, the equations could be $R = T/2$, $Y = T - R$, and $R + Y = T$.

Pages 37-42: Linking Circles

Strategies students can use: act out with objects, use a diagram, use logical thinking, and use a calculator.

Students use logical reasoning skills as they sort out the different sections and intersections of the linking circles, which are Venn diagrams. For example, in the first problem on page 37, students will find out that circle A has two sections—one not shared and one shared with circle B; circle B has three sections—one not shared, one shared with A, and one shared with C; and circle C has two sections—one not shared and one shared with circle B. Total values are given for a circle, which includes the intersection or intersections. Students will need to think about these problems, to understand how the intersections are subtracted from the totals to find values for a, b, and c.

Equations help to show how these problems are solved. To find the total, the equation for the first problem on page 37 is: $a + b + c + d + e =$ Total. Information is given in this problem for sections d and e, so the variables are sections a, b, and c. Equations for the variables are: $a = 7 - d$, $b = 7 - (d + e)$, and $c = 6 - e$.

Pages 43-48: What's the Rule?

Strategies students can use: make and use a table, use a diagram, look for a pattern, guess and check, act out with objects, and use a calculator.

In each problem, students are given pairs of input numbers for the first machine and output numbers for the second machine. There is a separate rule for each machine, so students will need to discover two rules for each problem, or the secret rule (function) for each machine. The numbers are given in a table. Students use cubes to act out the problem on diagrams of the machines. They can guess and check and also look for the pattern in the relationships between the numbers.

When students have found the rules and have described them in words, have them try to translate the rules into algebraic expressions. For example, the rules for the first problem on page 43 could be: $n + 2$; $n + 2 - 1$. The n stands for the first input number.

Solutions

These are sample solutions. Students may find other correct answers.

1

1. | Red | Yellow |
    |---|---|
    | 7 | 1 |
    | 6 | 2 |
    | 5 | 3 |

    $R > Y; R + Y = 8$

2. | Orange | Purple |
    |---|---|
    | 10 | 1 |
    | 9 | 2 |
    | 8 | 3 |
    | 7 | 4 |
    | 6 | 5 |

    $O > P; O + P = 11$

2

1. | White | Green |
    |---|---|
    | 13 | 1 |
    | 12 | 2 |
    | 11 | 3 |
    | 10 | 4 |
    | 9 | 5 |
    | 8 | 6 |

    $G < W; G + W = 14$

2. | Blue | Pink |
    |---|---|
    | 16 | 1 |
    | 15 | 2 |
    | 14 | 3 |
    | 13 | 4 |
    | 12 | 5 |
    | 11 | 6 |
    | 10 | 7 |
    | 9 | 8 |

    $P < B; B + P = 17$

3

1. | Green | Purple | Brown |
    |---|---|---|
    | 5 | 2 | 1 |
    | 4 | 3 | 1 |

    $G > P, P > B; G + P + B = 8$

2. | Orange | Blue | Yellow |
    |---|---|---|
    | 7 | 2 | 1 |
    | 6 | 3 | 1 |
    | 5 | 4 | 1 |
    | 5 | 3 | 2 |

    $O > B, B > Y; O + B + Y = 10$

4

1. | Blue | White | Red |
    |---|---|---|
    | 9 | 2 | 1 |
    | 8 | 3 | 1 |
    | 7 | 4 | 1 |
    | 6 | 5 | 1 |
    | 7 | 3 | 2 |
    | 6 | 4 | 2 |
    | 5 | 4 | 3 |

    $B > W, R < W; B + W + R = 12$

2. | Orange | Blue | Yellow |
    |---|---|---|
    | 12 | 2 | 1 |
    | 11 | 3 | 1 |
    | 10 | 4 | 1 |
    | 9 | 5 | 1 |
    | 8 | 6 | 1 |
    | 10 | 3 | 2 |
    | 9 | 4 | 2 |
    | 8 | 5 | 2 |
    | 7 | 6 | 2 |
    | 8 | 4 | 3 |
    | 7 | 5 | 3 |
    | 6 | 5 | 4 |

    $B < O, B > Y; O + B + Y = 15$

5

1. | Balloons | Horns |
    |---|---|
    | 29 | 1 |
    | 28 | 2 |
    | 27 | 3 |
    | 26 | 4 |
    | 25 | 5 |
    | 24 | 6 |
    | 23 | 7 |
    | 22 | 8 |
    | 21 | 9 |

    $B > 2H; B + H = 30$

2. | T-shirts | Sweaters |
    |---|---|
    | 49 | 1 |
    | 48 | 2 |
    | 47 | 3 |
    | 46 | 4 |
    | 45 | 5 |
    | 44 | 6 |
    | 43 | 7 |
    | 42 | 8 |

    $T > 5S; T + S = 50$

6

1. | Blue | Red |
    |---|---|
    | 24 | 1 |
    | 23 | 2 |
    | 22 | 3 |
    | 21 | 4 |
    | 20 | 5 |
    | 19 | 6 |

    $B > 3R; B + R = 25$

2. | C Chip | Oatmeal |
    |---|---|
    | 47 | 13 (12 + 1) |
    | 46 | 14 (12 + 2) |

    $C > 3O; C + O = 60$

3. | A's | Yankees | Giants |
    |---|---|---|
    | 23 | 13 | 4 |
    | 22 | 14 | 4 |
    | 21 | 15 | 4 |
    | 20 | 16 | 4 |
    | 19 | 17 | 4 |
    | 19 | 16 | 5 |
    | 18 | 17 | 5 |

    $Y > 3G, A > Y; Y + G + A = 40$

7

1. Carrots = 4, Nixes = 6, Potatoes = 9, Total = 19
   $C = 4, N = C + 2, P = N + 3; C + N + P =$ Total

2. Onions = 3, Miffles = 1, Tomatoes = 9, Total = 13
   $O = 3, M = O - 2, T = M + 8; O + M + T =$ Total

8

1. Tibbits = 6, Peppers = 15, Beets = 3, Total = 24
   $T = 6, P = T + 9, B = P - 12; T + P + B =$ Total

2. Zimmers = 5, Apples = 10, Peaches = 5, Total = 20
   $Z = 5, A = 2Z, P = A - 5; Z + A + P =$ Total

9

1. Onions = 4, Lizzers = 12, Potatoes = 9, Total = 25
   $O = 4, L = 3O, P = L - 3; O + L + P =$ Total

2. Mushrooms = 3, Zingles = 15, Tomatoes = 2, Total = 20
   $M = 3, Z = 5M, T = Z - 13; M + Z + T =$ Total

10

1. Zimmers = 4, Bananas = 7, Fizzits = 14, Oranges = 2, Total = 27
   $Z = 4, B = Z + 3, F = 2B, O = F - 12; Z + B + F + O =$ Total

2. Nixes = 3, Carrots = 9, Miffles = 3, Peppers = 13, Total = 28
   $N = 3, C = 3N, M = C - 6, P = M + 10; N + C + M + P =$ Total

11

1. Eggs = 4, Apples = 36, Brims = 83, Fizzits = 332, Peaches = 456, Total = 911
   $E = 4, A = 9E, B = A + 47, F = 4B, P = F + 124; E + A + B + F + P =$ Total

2. Eggs = 5, Brims = 29, Oranges = 203, Zimmers = 183, Bananas = 1098, Total = 1518
   $E = 5, B = E + 24, O = 7B, Z = O - 20, B = 6Z; E + B + O + Z + B =$ Total

12
1. Bushes = 20, Trees = 10, Vegetables = 40, Flowers = 120, Total = 190
$B = 20$, $T = B - 10$, $V = T + 30$, $F = 3V$;
$B + T + V + F =$ Total

2. Gold = 4, White = 6, Green = 12, Blue = 10, Black = 30, Total = 62
$Go = 4$, $W = Go + 2$, $Gr = 2W$, $Blu = Gr - 2$, $Bla = 3Blu$; $Go + W + Gr + Blu + Bla =$ Total

3. Calico = 5, White = 15, Gray = 12, Black & White = 24, Orange = 72, Total = 128
$C = 5$, $W = 3C$, $G = W - 3$, $B/W = 2G$, $O = 3B/W$; $C + W + G + B/W + O =$ Total

13
1.
⬡	◯	△
1	4	5

⬡ + ◯ = △
3⬡ + ◯ + △ = 12

2.
⬡	◯	△
1	6	7
2	4	6
3	2	5

⬡ + ◯ = △
3⬡ + ◯ + △ = 16

14
1.
▢	◯	△
1	7	2
2	5	4
4	1	8

2◯ = △
2◯ + 2▢ + △ = 18

2.
▢	◯	△
1	8	2
2	6	4
3	4	6
4	2	8

15
1.
◯	⬡	◇
2	3	1
3	1	5

2◯ − ⬡ = ◇
2◯ + 3⬡ + ◇ = 14

2.
◯	⬡	◇
3	4	2
4	2	6

16
1.
△	◯	⬡
3	5	1
4	3	5
5	1	9

2△ − ◯ = ⬡
2△ + 3◯ + ⬡ = 22

2.
△	◯	⬡
5	2	8

17
1.
◯	△	⬡
4	5	1
2	6	4

△ − ◯ = ⬡
3◯ + 3△ + ⬡ = 28

2.
◯	△	⬡
1	8	7
3	7	4
5	6	1

18
1.
Red	Green	Blue
1	13	2
2	11	4
3	9	6
4	7	8

$B = 2R$; $2R + 2G + B = 30$

2.
Apples	Peaches	Oranges
2	7	12
4	6	8
6	5	4

$2P - A = O$; $3A + 2P + O = 32$

3.
Box	Bottle	Ball
1	13	7
3	9	6

Bx Bx Bt —— Bx Ba Ba
Bt Ba —— Bx Bx Bx Ba

$2Ba - Bt = Bx$; $3Bx + 2Ba + Bt = 30$

19
1. 2 Zims and 5 Ziffs; 4 Zims and 4 Ziffs; 6 Zims and 3 Ziffs; 8 Zims and 2 Ziffs; 10 Zims and 1 Ziff (1 x *Zim*) + (2 x *Ziff*) = 12

2. 1 Zim and 8 Ziffs; 3 Zims and 7 Ziffs; 5 Zims and 6 Ziffs; 7 Zims and 5 Ziffs; 9 Zims and 4 Ziffs

20
1. 2 Dweeps and 4 Dwizzles; 5 Dweeps and 2 Dwizzles (2 x *Dwe*) + (3 x *Dwi*) = 16

2. 3 Dweeps and 6 Dwizzles; 6 Dweeps and 4 Dwizzles; 9 Dweeps and 2 Dwizzles

21
1. 1 Fip and 7 Fims; 3 Fips and 6 Fims; 5 Fips and 5 Fims; 7 Fips and 4 Fims; 9 Fips and 3 Fims; 11 Fips and 2 Fims; 13 Fips and 1 Fims (2 x *Fip*) + (4 x *Fim*) = 30

2. 1 Fip and 8 Fims; 3 Fips and 7 Fims; 5 Fips and 6 Fims; 7 Fips and 5 Fims; 9 Fips and 4 Fims; 11 Fips and 3 Fims; 13 Fips and 2 Fims

22
1. 1 Snoof and 5 Snuffles; 6 Snoofs and 2 Snuffles (3 x *Sno*) + (5 x *Snu*) = 28

2. 5 Snoofs and 5 Snuffles; 10 Snoofs and 2 Snuffles

23
1. 13 Huffs and 11 Hissles (5 x *Hu*) + (6 x *Hi*) = 131

2. 15 Huffs and 17 Hissles

24
1. No, Nancy's school did not win! 17 boxes of bottles (340) plus 9 boxes of paper (315) is 26 boxes for a total of 655 coupons.
$20B + 35P = 655$

2. 11 walking miles ($3.30) plus 18 running miles ($8.10) is 29 miles for a total of $11.40.
$0.30W + 0.45R = 11.40$

3. They took off a box of Crumbles. 5 boxes of Flings (25 lbs.) plus 9 boxes of Crumbles (36 lbs.) is 14 boxes for a total of 61 pounds.
$5F + 4C = 61$

25 1.

Day	Fizzywits	Moosles
1	2	5
2	4	9
3	6	13
4	8	17
5	10	21

Number of Fizzywits increases by 2, Moosles by 4.
$F = 2n, M = 4n + 1$

2. 8 Bizzies, 19 Purples
Number of Bizzies increases by 1, Purples by 3.
$B = n + 2, P = 3n + 1$

26 1

Day	Fiddles	Gillies
1	9	16
2	8	13
3	7	10
4	6	7
5	5	4
6	4	1

Number of Fiddles decreases by 1, Gillies by 3.
$F = 10 - n, G = 19 - 3n$

2. 4 Scarlets, 8 Leopards
Number of Scarlets decreases by 2, Leopards by 4.
$S = 14 - 2n, L = 28 - 4n$

27 1

Day	Barrels	Floofies
1	1	11
2	6	10
3	11	9
4	16	8
5	21	7

Number of Barrels increases by 5,
Floofies decrease by 1.
$B = 5n - 4, F = 12 - n$

2. 4 Hog Noses, 17 Silvers
Number of Hog Noses decreases by 3,
Silvers increase by 3.
$H = 22 - 3n, S = 3n - 1$

28 1.

Day	Metallica	Mugo	Fruticosa
1	22	5	7
2	18	7	6
3	14	9	5
4	10	11	4
5	6	13	3

Number of Metallicas decreases by 4,
Mugos increase by 2,
Fruticosas decrease by 1.
$Me = 26 - 4n, Mu = 4n + 1, F = 8 - n$

2. 31 Pinks, 3 Spotteds, 23 Bufos
Number of Pinks increases by 6,
Spotteds decrease by 5, Bufos increase by 4.
$P = 6n - 5, S = 33 - 5n, B = 4n - 1$

29 1.

Day	Trilobas	Dumosas	Lobatas
1	25	120	55
2	45	108	64
3	65	96	73
4	85	84	82
⋮	⋮	⋮	⋮
10	205	12	136

Number of Trilobas increases by 20,
Dumosas decrease by 12, Lobatas increase by 9.
$T = 20n + 5, D = 132 - 12n, L = 9n + 46$

2. 97 Sillies, 60 Bananas, and 180 Fuzzies
Number of Sillies increases by 8,
Bananas decrease by 10, Fuzzies increase by 15.
$S = 8n + 17, B = 160 - 10n, F = 15n + 30$

30 1. 3:00, rule for the pattern: $12n - 8$
2. Bag 11, rule for pattern: $315 - 15n$
3. 7th hour, rules for pattern: $B = 6n + 8, S = 5n + 15$

31 1. Yellow = 6, Total = 12
$R = T/2, Y = T - R; R + Y = T$

2. Pink = 4, Total = 16
$B = 3/4T, P = T - B; B + P = T$

32 1. Purple = 12, Total = 24
$G = T/6, O = T/3, P = T - (G + O); O + P + G = T$

2. White = 10, Total = 30 $B = T/2, R = T/6,$
$W = T - (B + R); R + B + W = T$

33 1. Yellow = 10, Total = 16
$O = T/8, B = T/4, Y = T - (O + B); B + O + Y = T$

2. Yellow = 9, Total = 24
$O = 3/8T, B = T/4, Y = T - (O + B); B + O + Y = T$

34 1. Brown = 9, Total = 18
$Y = T/6, G = T/3, B = T - (Y + G); B + G + Y = T$

2. Green = 10, Total = 24 $P = 3/12T, W = T/3, G = T - (P + W); W + P + G = T$

35 1. Basketball = 20, Total = 120
$Te = T/12, Base = T/4, S = T/2, Basket = T - (Te + Base + S); Base + Basket + S + Te = T$

2. Tennis = 5, Total = 80
$S = 5/16T, F = 3/8T, B = T/4, Te = T - (S + F + B); B + Te + S + F = T$

36 1. Orange = 6, Total = 48
$B = T/8, Pu = T/4, Pi = T/2, O = T - (B + Pu + Pi); Pi + Pu + B + O = T$

2. Hamburger = 4, Total = 48
$Sa = T/12, V = T/12, Super = 5/12T, Supreme = T/3, H = T - (Sa + V + Super + Supreme); Supreme + Super + V + Sa + H = T$

3. Massachussetts = 4, Total = 32
$MI = T/16$, $NY = 3/16 T$, $CA = 3/8 T$, $TX = T/4$, $MA = T - (MI + NY + CA + TX)$; $CA + TX + NY + MI + MA = T$

37 1. Total = 15 coins 2. Total = 24 cards

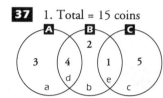 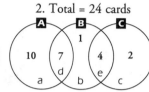

For 1: $a = 7 - d$, $b = 7 - (d + e)$, $c = 6 - e$; $a + b + c + d + e$ = Total

38 1. Total = 25 beads 2. Total = 33 hats

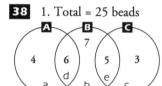 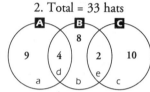

39 1. Total = 20 beads 2. Total = 29 coins

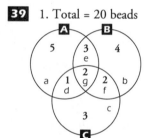 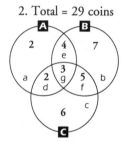

For 1: $a = 11 - (d + g + e)$, $b = 11 - (e + g + f)$, $c = 8 - (d + g + f)$, $d = 3 - g$, $e = 5 - g$, $f = 4 - g$; $a + b + c + d + e + f + g$ = Total

40 1. Total = 38 cards 2. Total = 52 hats

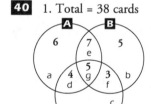 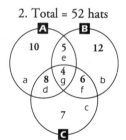

41 1. Total = 123 charms 2. Total = 429 coins

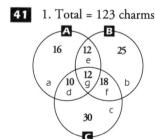 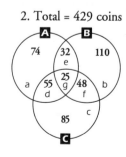

42 1. Total = 24 trucks

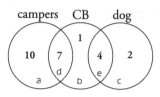

2. Total = 25 pairs of socks   3. Total = 58 people

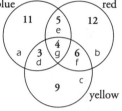 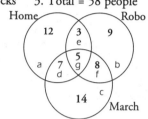

For pages 43 through 47, students may discover other rules.

43 1. $n + 2$; $n + 2 - 1$ 2. $n - 1$; $n - 1 + 5$
44 1. $n + 8$; $n + 8 - 3$ 2. $2n$; $2n + 4$
45 1. $n + 6$; $(n + 6) \times 2$ 2. $3n$; $3n - 3$
46 1. $n + 12$; $n + 12 - 6$ 2. $4n$; $4n - 10$
47 1. $5n$; $5n + 25$ 2. $30n$; $30n + 4$
48 1. 80 Wudgets $n + 5$; $2(n + 5)$
2. $259 $3n$; $3n + 4$
3. The Goof would send back 57 balls. $n - 6$; $3(n - 6)$

49 1. 158 bottles   2.

△	□	○
1	10	11
2	8	10
3	6	9
5	2	7

3. △ + ○ + □ = 24

3. No, because she hit 8 small bears and 14 large bears.

50 1. 5 hours   2.

White	Yellow
64	11
63	12
62	13
61	14
60	15

Pattern rules: $D = 7n + 8$
$J = 9n + 1$

3. Raccoons 15, Total = 96

51 1. 164 treasures   2. 108 people
3. 29 large cars, 17 small cars

52 1. Yes   2. Cats - 112, dogs - 56, fish - 14, birds - 9, rabbits - 6
Pattern rules:
$L = 2n + 3$, $A = 3n - 1$,
$K = 16 - n$

3. Purple - 10.   Red and gold were colors chosen.

53 1.

Purple	Red	Blue
32	13	10
31	14	10
30	15	10
30	14	11

2. 31, rule: $n \div 2$; $(n \div 2) + 8$

3.

Chocolate	Lemon	Cinnamon
5	8	2
7	4	10
8	2	14

54 1. 66 donuts would come out. Rule: $n - 5$; $6(n - 5)$
2. 146 students
3. Poetry - 6, 120 people voted.

Beginning Algebra Thinking, Grades 5-6 • ©Ideal School Supply Company

Number Combinations - 1

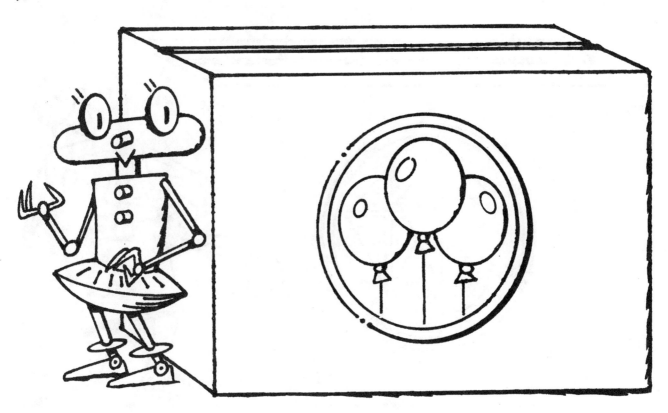

Use cubes to solve each problem.

1 There are eight balloons in the box—some red, some yellow. There are more red balloons than yellow balloons. How many balloons of each color could be in the box?

Red	Yellow

2 There are 11 balloons in the box—some orange, some purple. There are more orange balloons than purple balloons. How many balloons of each color could be in the box?

Orange	Purple

Beginning Algebra Thinking, Grades 5-6 • ©Ideal School Supply Company

Number Combinations - 2

Use cubes to solve each problem.

1 There are 14 kites in the box—some white, some green.
There are fewer green kites than white kites.
How many kites of each color could be in the box?

White	Green

2 There are 17 kites in the box—some pink, some blue.
There are fewer pink kites than blue kites.
How many kites of each color could be in the box?

Blue	Pink

Beginning Algebra Thinking, Grades 5-6 • ©Ideal School Supply Company

Number Combinations - 3

Use cubes to solve each problem.

1 There are eight T-shirts in the box—some brown, some purple, some green. There are more purple shirts than brown shirts. There are more green shirts than purple shirts. How many shirts of each color could be in the box?

Green	Purple	Brown

2 There are 10 T-shirts in the box—some yellow, some orange, some blue. There are more blue shirts than yellow shirts. There are more orange shirts than blue shirts. How many shirts of each color could be in the box?

Orange	Blue	Yellow

Beginning Algebra Thinking, Grades 5-6 • ©Ideal School Supply Company

Number Combinations - 4

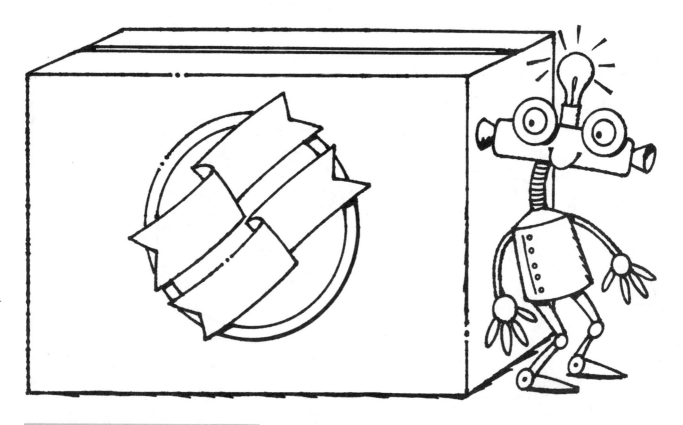

Use cubes to solve each problem.

1 There are 12 banners in the box—some blue, some white, some red. There are more blue banners than white banners. There are fewer red banners than white banners. How many banners of each color could be in the box?

Blue	White	Red

2 There are 15 banners in the box—some yellow, some black, some orange. There are fewer black banners than orange banners. There are more black banners than yellow banners. How many banners of each color could be in the box?

Orange	Black	Yellow

Number Combinations - 5

Use a calculator to solve these problems.

1 There are 30 balloons and horns in the box.
There are more than twice as many balloons as horns.
How many balloons and how many horns could be in the box?

Balloons	Horns

2 There are 50 T-shirts and sweaters in the box.
There are more than five times as many T-shirts as sweaters.
How many T-shirts and how many sweaters could be in the box?

T-shirts	Sweaters

Number Combinations - 6

1 Annelise calls Chris to come and take a look at the new fish tank. There are 25 fish in the tank—some blue and some red. There are more than three times as many blue fish as red fish. How many fish of each color could be in the tank?

2 Marietta is filling the display case at the Cookie Factory. When she begins, there are a dozen oatmeal cookies in the case. She adds chocolate chip cookies and oatmeal cookies until there are 60 cookies in all. There are more than three times as many chocolate chip cookies as oatmeal cookies. How many chocolate chip cookies and how many oatmeal cookies could be in the case?

3 In Marco's dream, monsters wearing baseball caps were chasing him through the woods. There were at least four Giants caps. There were more than three times as many Yankees caps as Giants caps. Then there were more As caps than Yankees caps. There were 40 monsters running after Marco! How many baseball caps of each kind could the monsters have been wearing?

Number Puzzles - 1

Use cubes to solve each problem.

1 Tron put potatoes, nixes, and carrots into the space blender. He put in three more potatoes than nixes and two more nixes than carrots. He put in four carrots. How many of each thing did Tron put in? How many things in all?

Carrots _____ Nixes _____ Potatoes _____ Total _____

2 Thara dropped tomatoes, miffles, and onions into the space blender. She put in eight more tomatoes than miffles and two fewer miffles than onions. She put in three onions. How many of each thing did Thara put in? How many things in all?

Onions _____ Miffles _____ Tomatoes _____ Total _____

Beginning Algebra Thinking, Grades 5-6 • ©Ideal School Supply Company

Number Puzzles - 2

Use cubes to solve each problem.

1 Thara mixed peppers, tibbits, and beets in the space blender. She put in 12 fewer beets than peppers and nine more peppers than tibbits. She put in six tibbits. How many of each thing did Thara put in? How many things in all?

Tibbits _____ Peppers _____ Beets _____ Total _____

2 Tron mixed apples, zimmers, and peaches in the space blender. He put in five fewer peaches than apples and two times as many apples as zimmers. He put in five zimmers. How many of each thing did Tron put in? How many things in all?

Zimmers _____ Apples _____ Peaches _____ Total _____

Number Puzzles - 3

Use cubes to solve each problem.

1 Tron put onions, lizzers, and potatoes into the space blender. He put in three fewer potatoes than lizzers and three times as many lizzers as onions. He put in four onions. How many of each thing did he put in? How many things in all?

Onions _____   Lizzers _____   Potatoes _____   Total _____

2 Thara added mushrooms, zingles, and tomatoes to the space blender. She put in 13 fewer tomatoes than zingles and five times as many zingles as mushrooms. She put in three mushrooms. How many of each thing did Thara put in? How many things in all?

Mushrooms _____   Zingles _____   Tomatoes _____   Total _____

Beginning Algebra Thinking, Grades 5-6 • ©Ideal School Supply Company

Number Puzzles - 4

Use cubes to solve each problem.

1 Tron put fizzits, oranges, zimmers, and bananas into the space blender. He put in 12 fewer oranges than fizzits and two times as many fizzits as bananas. He put in three more bananas than zimmers. He put in four zimmers. How many of each thing did Tron put in? How many things in all?

Zimmers _____   Bananas _____   Fizzits _____   Oranges _____   Total _____

2 Thara put peppers, miffles, nixes, and carrots into the space blender. She put in ten more peppers than miffles and six fewer miffles than carrots. She put in three times as many carrots as nixes. She put in three nixes. How many of each thing did Thara put in? How many things in all?

Nixes _____   Carrots _____   Miffles _____   Peppers _____   Total _____

Number Puzzles - 5

Use a calculator to solve each problem.

1 Thara made a Super Space Shake, using the space blender. She put in 124 more peaches than fizzits. She put in four times as many fizzits as brims and 47 more brims than apples. She put in nine times as many apples as eggs. She put in four eggs. How many of each thing did Thara put in? How many things in all?

Eggs _____ Apples _____ Brims _____ Fizzits _____ Peaches _____ Total _____

2 Tron made a Super Space Shake, using the space blender. He put in six times as many bananas as zimmers. He put in 20 fewer zimmers than oranges and seven times as many oranges as brims. He put in 24 more brims than eggs. He put in five eggs. How many of each thing did Tron put in? How many things in all?

Eggs _____ Brims _____ Oranges _____ Zimmers _____ Bananas _____ Total _____

Beginning Algebra Thinking, Grades 5-6 • ©Ideal School Supply Company

Number Puzzles - 6

1 Beryl works after school and Saturdays at Garden Mart. Today she watered the plants just before closing time. She watered three times as many flowers as vegetable plants, 30 more vegetable plants than trees, 10 fewer trees than bushes, and 20 bushes. How many of each thing did she water? How many in all?

2 MegaMart is having a Mega Sale! Darius was counting shirts for the sale. He counted three times as many black shirts as blue shirts, two fewer blue shirts than green shirts, two times as many green shirts as white shirts, two more white shirts than gold shirts, and four gold shirts. How many of each color shirt did Darius count? How many in all?

3 The City Cat Show is under way, complete with purring, meowing, and an occasional hiss. There are three times as many orange cats as black and white cats, two times as many black and white cats as gray cats, three fewer gray cats than white cats, three times as many white cats as calico cats, and five calico cats. How many of each kind of cat is in the show? How many in all?

12   Beginning Algebra Thinking, Grades 5-6 • ©Ideal School Supply Company

Seesaw Balances - 1

> Use cubes to solve the problems.
> Find as many solutions as you can.
>
> Rules:
> Balance the seesaw.
> Robots that are the same have the same weight.
> Robots that are different have different weights.
> All robots weigh more than zero pounds.

1 If the robots on the seesaw weigh 12 pounds all together, what could each kind of robot weigh?

2 If the robots on the seesaw weigh 16 pounds all together, what could each kind of robot weigh?

Beginning Algebra Thinking, Grades 5-6 • ©Ideal School Supply Company

Seesaw Balances - 2

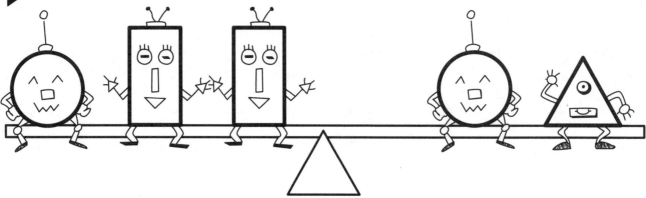

Use cubes to solve the problems.
Find as many solutions as you can.

Rules:
Balance the seesaw.
Robots that are the same have the same weight.
Robots that are different have different weights.
All robots weigh more than zero pounds.

1 If the robots on the seesaw weigh 18 pounds all together, what could each kind of robot weigh?

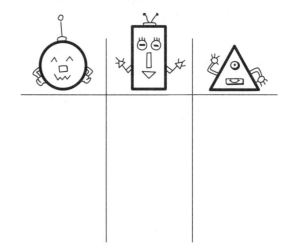

2 If the robots on the seesaw weigh 20 pounds all together, what could each kind of robot weigh?

Seesaw Balances - 3

Use cubes to solve the problems.
Find as many solutions as you can.

Rules:
Balance the seesaw.
Robots that are the same have the same weight.
Robots that are different have different weights.
All robots weigh more than zero pounds.

1 If the robots on the seesaw weigh 14 pounds all together, what could each kind of robot weigh?

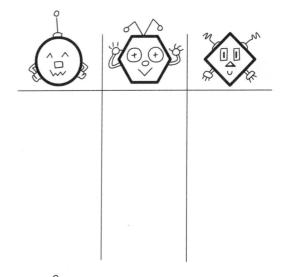

2 If the robots on the seesaw weigh 20 pounds all together, what could each kind of robot weigh?

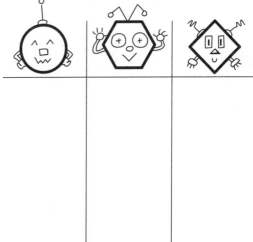

Beginning Algebra Thinking, Grades 5-6 • ©Ideal School Supply Company

Seesaw Balances - 4

Use cubes to solve the problems.
Find as many solutions as you can.

Rules:
Balance the seesaw.
Robots that are the same have the same weight.
Robots that are different have different weights.
All robots weigh more than zero pounds.

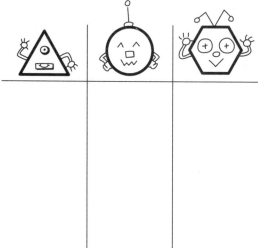

1 If the robots on the seesaw weigh 22 pounds all together, what could each kind of robot weigh?

2 If the robots on the seesaw weigh 24 pounds all together, what could each kind of robot weigh?

Beginning Algebra Thinking, Grades 5-6 • ©Ideal School Supply Company

Seesaw Balances - 5

Use a calculator to solve the problems.
Find as many solutions as you can.

Rules:
Balance the seesaw.
Robots that are the same have the same weight.
Robots that are different have different weights.
All robots weigh more than zero pounds.

1 If the robots on the seesaw weigh 28 pounds all together, what could each kind of robot weigh?

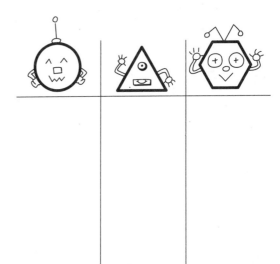

2 If the robots on the seesaw weigh 34 pounds all together, what could each kind of robot weigh?

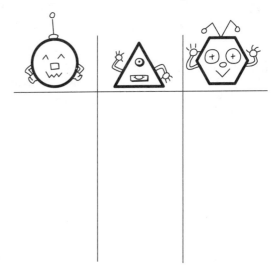

Beginning Algebra Thinking, Grades 5-6 • ©Ideal School Supply Company

Seesaw Balances - 6

1 It's time for "Guess The Secret Numbers." Tonight there are two red doors and one green door on the left of the MC, and one green door and one blue door on the right of the MC. Shirelle, the MC, gives directions to the contestants: "Each door hides a number. Doors that are the same color hide the same numbers. Doors that are different colors hide different numbers. Every number is greater than zero. The sum of the numbers on each side of me is the same and the total sum is 30. What are the secret numbers tonight?"

2 Nicole and James sorted the bags so that they could each take home the same amount of fruit. Each bag of apples had the same number of apples in it. Each bag of peaches had the same number of peaches. There was only one bag of oranges. Each kind of fruit had a different number of pieces. All together there were 32 pieces of fruit. James took home two bags of apples and one bag of oranges. Nicole took home one bag of apples and two bags of peaches. How many pieces of fruit could have been in each bag?

3 Marvelous Marlena is preparing for her high wire act featuring a glow-in-the-dark balancing pole. She has three sparkling boxes, two shiny balls, and one shimmering bottle. Together the things weigh 30 ounces. All boxes weigh the same and all balls weigh the same. Boxes, balls, and bottles weigh different numbers of ounces. She attaches the boxes, balls, and bottle to the two ends of her pole to make it balance. What could the weight of each box, ball, and bottle be? How could she attach them?

Ratios and Number Combinations - 1

Fill in the table.

Use cubes and the table to solve the problems.

1 If there are 12 in the space ship, how many Zims and how many Ziffs could be in the ship?

2 If there are 17 in the space ship, how many Zims and how many Ziffs could be in the ship?

Number of Zims	Number of Antennas	Number of Ziffs	Number of Antennas
1	1	1	2
2	2	2	4
		3	6

Beginning Algebra Thinking, Grades 5-6 • ©Ideal School Supply Company

Ratios and Number Combinations - 2

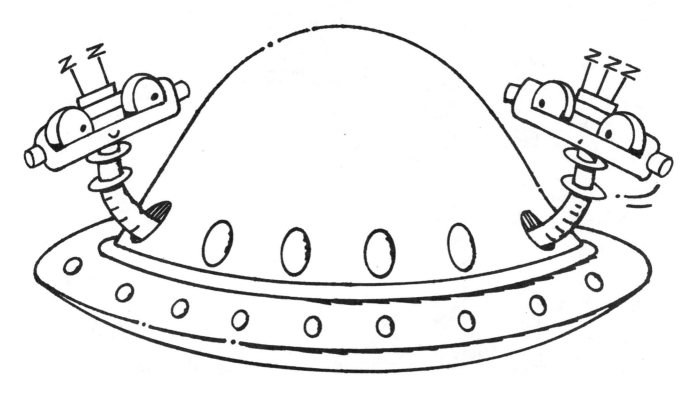

Fill in the table.
Use cubes and the table to solve the problems.

Number of Dweeps	Number of Antennas	Number of Dwizzles	Number of Antennas

1 If there are 16 Z in the space ship, how many Dweeps and how many Dwizzles could be in the ship?

2 If there are 24 Z in the space ship, how many Dweeps and how many Dwizzles could be in the ship?

Beginning Algebra Thinking, Grades 5-6 • ©Ideal School Supply Company

Ratios and Number Combinations - 3

Fill in the table.
Use cubes and the table to solve the problems.

Number of Fips	Number of Antennas	Number of Fims	Number of Antennas

1 If there are 30 antennas in the space ship, how many Fips and how many Fims could be in the ship?

2 If there are 34 antennas in the space ship, how many Fips and how many Fims could be in the ship?

Beginning Algebra Thinking, Grades 5-6 • ©Ideal School Supply Company

Ratios and Number Combinations - 4

Fill in the table.
Use cubes and the table to solve the problems.

1 If there are 28 in the space ship, how many Snoofs and how many Snuffles could be in the ship?

2 If there are 40 in the space ship, how many Snoofs and how many Snuffles could be in the ship?

Number of Snoofs	Number of Antennas	Number of Snuffles	Number of Antennas

Beginning Algebra Thinking, Grades 5-6 • ©Ideal School Supply Company

Ratios and Number Combinations - 5

Fill in the table.

Use a calculator and the table to solve the problems.

 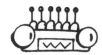

Number of Huffs	Number of Antennas	Number of Hissles	Number of Antennas

1 If there are 24 creatures all together in the space ship and 131 ↑, how many Huffs and how many Hissles are in the ship?

2 If there are 32 creatures all together in the space ship and 177 ↑, how many Huffs and how many Hissles are in the ship?

Beginning Algebra Thinking, Grades 5-6 • ©Ideal School Supply Company

Ratios and Number Combinations - 6

1 The schools in Nancy's district are collecting bottles and paper to turn in for coupons. Each box of bottles is worth 20 coupons, and each box of paper is worth 35 coupons. The first school to get 350 paper coupons or 400 bottle coupons wins a big prize. Nancy's school has 655 coupons and the students have collected a total of 26 boxes. Did they win a prize?

2 It was the annual Turkey Walk and Run. Sponsers paid $0.30 for each mile that someone walked and $0.45 for each mile that someone ran. Together Adriana and her family walked and ran a total of 29 miles. Each person who sponsored the family paid $11.40. How many miles did Adriana's family run and how many miles did they walk?

3 Sarah and Molly packed the little plane with supplies for Manka. First they packed the essentials. Then they packed 14 boxes of snack food—some boxes of Flings and some of Crumbles. Each box of Flings weighed five pounds and each box of Crumbles weighed four pounds. When they discovered that the plane was over the weight limit by three pounds, they took off a snack food box. If the total weight of the snack food boxes was 61 pounds, and they took a box from the group having the most boxes, did they take off a box of Flings or a box of Crumbles?

Beginning Algebra Thinking, Grades 5-6 • ©Ideal School Supply Company

Number Patterns - 1

Use cubes to solve the problems.

1 Biologist Zingbat is looking for strange plants. On the first day he finds 2 fizzywits and 5 moosles. In the second day he spots 4 fizzywits and 9 moosles. On the third day he sees 6 fizzywits and 13 moosles. On the fourth day he finds 8 fizzywits and 17 moosles. If these patterns continue, how many fizzywits and moosles will he find on the fifth day?

Day	Fizzywits	Moosles
1	2	5
2	4	
3		

2 Zoologist Whizzit is looking for exotic bugs. On the first day she discovers 3 bizzies and 4 rare purples. On the second day she spots 4 bizzies and 7 purples. On the third day she finds 5 bizzies and 10 purples. On the fourth day she sees 6 bizzies and 13 purples. If these patterns continue, how many bizzies and purples will she find on the sixth day?

Day	Bizzies	Purples

Beginning Algebra Thinking, Grades 5-6 • ©Ideal School Supply Company

Number Patterns - 2

Use cubes to solve the problems.

1 Biologist Zingbat is looking for rare trees. On the first day he spots 9 fiddles and 16 gillies. On the second day he discovers 8 fiddles and 13 gillies. On the third day he finds 7 fiddles and 10 gillies. On the fourth day he sees 6 fiddles and 7 gillies. If these patterns continue, how many fiddles and gillies will he see on the sixth day?

Day	Fiddles	Gillies

2 Zoologist Whizzit is searching for interesting frogs. On the first day she finds 12 scarlets and 24 leopards. On the second day she discovers 10 scarlets and 20 leopards. On the third day she spots 8 scarlets and 16 leopards. On the fourth day she sees 6 scarlets and 12 leopards. If these patterns continue, how many scarlets and leopards will she see on the fifth day?

Day	Scarlets	Leopards

Number Patterns - 3

Use cubes to solve the problems.

1 Biologist Zingbat is looking for rare bushes. On the first day he spots 1 barrel and 11 floofies. On the second day he discovers 6 barrels and 10 floofies. On the third day he finds 11 barrels and 9 floofies. On the fourth day he spies 16 barrels and 8 floofies. If these patterns continue, how many barrels and floofies will he find on the fifth day?

Day	Barrels	Floofies

2 Zoologist Whizzit is looking for interesting snakes. On the first day she finds 19 hog noses and 2 silvers. On the second day she sees 16 hog noses and 5 silvers. On the third day she spots 13 hog noses and 8 silvers. On the fourth day she discovers 10 hog noses and 11 silvers. If these patterns continue, how many hog noses and silvers will she spot on the sixth day?

Day	Hog Noses	Silvers

Beginning Algebra Thinking, Grades 5-6 • ©Ideal School Supply Company

Number Patterns - 4

Use cubes to solve the problems.

1 Biologist Zingbat is searching for rare vines. On the first day he discovers 22 metallicas, 5 mugos, and 7 fruticosas. On the second day he uncovers 18 metallicas, 7 mugos, and 6 fruticosas. On the third day he spots 14 metallicas, 9 mugos, and 5 fruticosas. On the fourth day he finds 10 metallicas, 11 mugos, and 4 fruticosas. If these patterns continue, how many metallicas, mugos, and fruticosas will he find on the fifth day?

Day	Metallicas	Mugos	Fruticosas

2 Zoologist Whizzit is looking for exotic toads. On the first day she spots 1 pink, 28 spotteds, and 3 bufos. On the second day she discovers 7 pinks, 23 spotteds, and 7 bufos. On the third day she finds 13 pinks, 18 spotteds, and 11 bufos. On the fourth day she sees 19 pinks, 13 spotteds, and 15 bufos. If these patterns continue, how many pinks, spotteds, and bufos will she find on the sixth day?

Day	Pinks	Spotteds	Bufos

Beginning Algebra Thinking, Grades 5-6 • ©Ideal School Supply Company

Number Patterns - 5

Use a calculator to solve the problems.

1 Biologist Zingbat is searching for rare plants. On the first day he finds 25 trilobas, 120 dumosas, and 55 lobatas. On the second day he discovers 45 trilobas, 108 dumosas, and 64 lobatas. On the third day he sees 65 trilobas, 96 dumosas, and 73 lobatas. On the fourth day he spots 85 trilobas, 84 dumosas, and 82 lobatas. If these patterns continue, how many trilobas, dumosas, and lobatas will he find on the tenth day?

Day	Trilobas	Dumosas	Lobatas

2 Zoologist Whizzit is looking for exotic slugs. On the first day she uncovers 25 sillies, 150 bananas, and 45 fuzzies. On the second day she sees 33 sillies, 140 bananas, and 60 fuzzies. On the third day she spots 41 sillies, 130 bananas, and 75 fuzzies. On the fourth day she finds 49 sillies, 120 bananas, and 90 fuzzies. If these patterns continue, how many sillies, bananas, and fuzzies will she see on the tenth day?

Day	Sillies	Bananas	Fuzzies

Beginning Algebra Thinking, Grades 5-6 • ©Ideal School Supply Company

Number Patterns - 6

1 It is 2:00 and the tickets are going on sale for Big Bug's Concert. In the first ten minutes 4 people get in line. During the next ten minutes 16 people get in line. During the third ten minutes 28 people get in line. In the fourth ten minutes 40 people get in line. If people keep lining up in this way, at what time will there be more than 200 people all together in line?

2 Dan and Mark were on a boat hunting for pirate treasure. One day they pulled up a big trunk with lots of numbered bags in it. Bag 1 had 300 coins in it, bag 2 had 285 coins, bag 3 had 270 coins, and bag 4 had 255 coins. If this pattern continued, in what bag did they find 150 coins?

3 It's a hectic Saturday! Benny and Sue are making burgers as fast as they can at Burger City. In the first hour Benny makes 14 burgers and Sue makes 20. In the second hour Benny makes 20 burgers and Sue makes 25. In the third hour Benny makes 26 and Sue makes 30. In the fourth hour Benny makes 32 and Sue makes 35. If they continue making burgers at those rates, during what hour will they make the same number of burgers?

Fraction Puzzles - 1

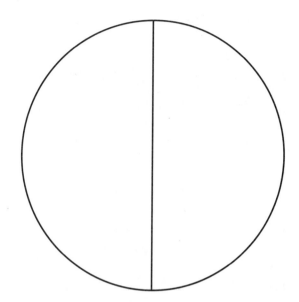
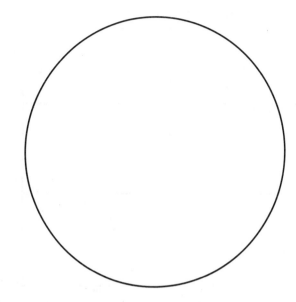

**Use cubes to solve the problems.
Make a diagram for each problem.**

1 Max planted red and yellow roses in a circle. One-half of the roses were red. There were six red roses.

　　　How many roses were yellow? _____

　　　How many roses did Max plant? _____

2 Carmen planted blue and pink tulips in a circle. Three-fourths of the tulips were blue. There were 12 blue tulips.

　　　How many tulips were pink? _____

　　　How many tulips did Carmen plant? _____

Beginning Algebra Thinking, Grades 5-6 • ©Ideal School Supply Company

Fraction Puzzles - 2

Use cubes to solve the problems.
Make a diagram for each problem.

1 Bill packed orange, purple, and green horns in a box. One-third of the horns were orange. One-sixth of the horns were green. Four of the horns were green.

How many horns were purple? _____

How many horns did Bill pack? _____

2 Marta packed red, blue, and white party hats in a box. One-half of the hats were blue. One-sixth of the hats were red. There were five red hats.

How many hats were white? _____

How many hats did Marta pack? _____

Fraction Puzzles - 3

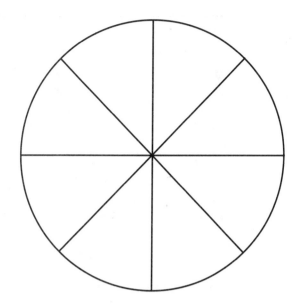
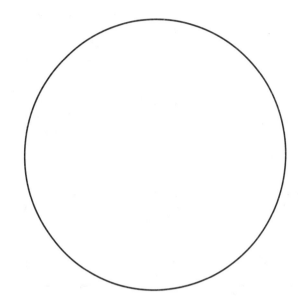

Use cubes to solve the problems.
Make a diagram for each problem.

1 Rosario put black, orange, and yellow candy witches in the bowl. One-fourth of the witches were black. One-eighth of the witches were orange. There were two orange witches.

How many witches were yellow? _____

How many witches were in the bowl? _____

2 Arnold put black, orange, and yellow candy cats in the bowl. One-fourth of the cats were black. Three-eighths of the cats were orange. There were nine orange cats.

How many cats were yellow? _____

How many cats were in the bowl? _____

Beginning Algebra Thinking, Grades 5-6 • ©Ideal School Supply Company

Fraction Puzzles - 4

Use cubes to solve the problems.
Make a diagram for each problem.

1 Rodney put brown, green, and yellow caps in the display case. One-third of the caps were green. One-sixth of the caps were yellow. There were three yellow caps.

How many caps were brown? _____

How many caps were in the case? _____

2 Becky put white, pink, and green T-shirts in the display case. One-third of the shirts were white. Three-twelfths of the shirts were pink. There were six pink shirts.

How many shirts were green? _____

How many shirts were in the case? _____

Fraction Puzzles - 5

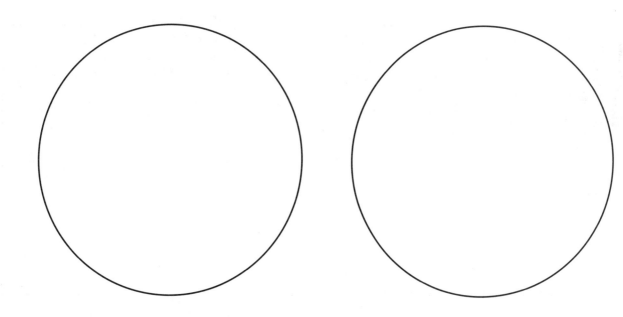

Use a calculator to solve the problems.
Make a diagram for each problem.

1 On Saturday, the park was filled with people playing baseball, basketball, soccer, and tennis. One-half of the people were playing soccer, one-fourth were playing baseball, and one-twelfth were playing tennis. There were 10 people playing tennis.

How many people were playing basketball? _____

How many people were at the park? _____

2 On Sunday, the park was filled with people playing baseball, tennis, soccer, and football. One-fourth of the people were playing baseball, three-eighths were playing football, and five-sixteenths were playing soccer. There were 25 people playing soccer.

How many people were playing tennis? _____

How many people were at the field? _____

Beginning Algebra Thinking, Grades 5-6 • ©Ideal School Supply Company

Fraction Puzzles - 6

1 It's Saturday morning, and Melanie just unpacked pink, purple, blue, and orange sweaters at Warm Fuzzies. She discovered that one-half of the new sweaters were pink, one-fourth of them were purple, and one-eighth of them were blue. There were six blue sweaters. How many sweaters were orange? How many sweaters did Melanie unpack?

2 Debbie is working part-time at Pizza on Wheels in CitySphere, the city in a bubble. She delivers supreme, super, veggie, sausage, and hamburger pizzas on her bicycle. Last week one-third of her deliveries were supreme, five-twelfths were super, one-twelfth were veggie, and one-twelfth were the sausage. She delivered four sausage pizzas. How many hamburger pizzas did Debbie deliver? How many pizzas did she deliver all together?

3 There were cooks from Texas, California, New York, Massachusetts, and Michigan at the annual Greatest Burger Cookoff. One-fourth of the cooks came from Texas, three-eighths from California, three-sixteenths from New York, and one-sixteenth from Michigan. There were two cooks from Michigan. How many cooks came from Massachusetts? How many cooks were there all together?

Linking Circles - 1

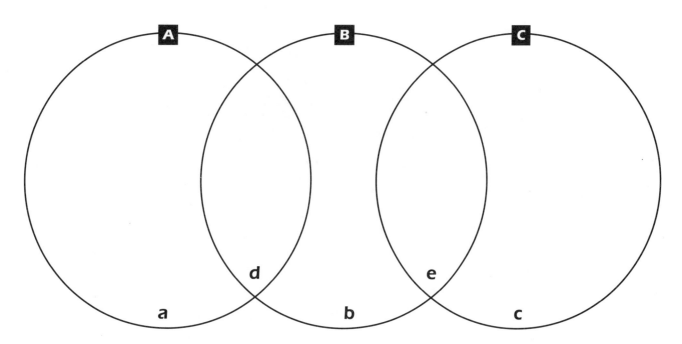

Use cubes to solve the problems.

1 The magician dropped coins into the linking circles. There are seven coins in circle A, seven coins in circle B, and six coins in circle C. Four of the coins are in both circles A and B. One of the coins is in both circles B and C. How many coins did the magician drop into the circles?

2 The magician dropped cards into the linking circles. There are 17 cards in circle A, 12 cards in circle B, and six cards in circle C. Seven of the cards are in both circles A and B. Four of the cards are in both circles B and C. How many cards did the magician drop into the circles?

Beginning Algebra Thinking, Grades 5-6 • ©Ideal School Supply Company

Linking Circles - 2

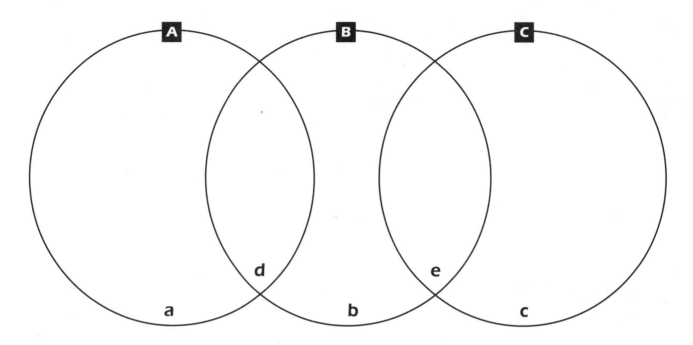

Use cubes to solve the problems.

1 The magician dropped beads into the linking circles. There are 10 beads in circle A, 18 beads in circle B, and eight beads in circle C. Six of the beads are in both circles A and B. Five of the beads are in both circles B and C. How many beads did the magician drop into the circles?

2 The magician dropped hats into the linking circles. There are 13 hats in circle A, 14 hats in circle B, and 12 hats in circle C. Four of the hats are in both circles A and B. Two of the hats are in both circles B and C. How many hats did the magician drop into the circles?

Linking Circles - 3

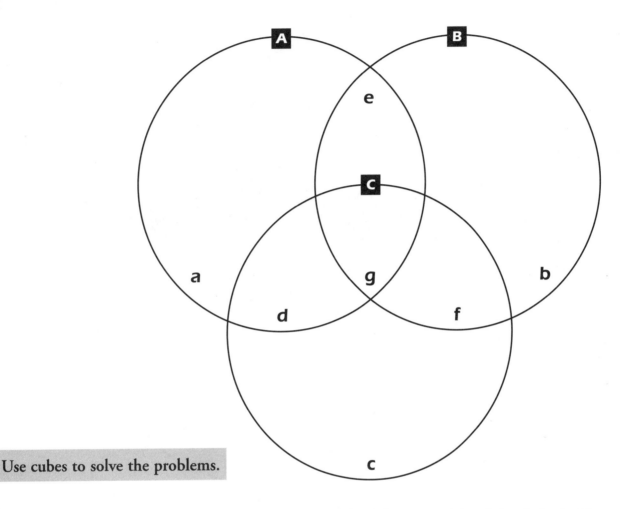

Use cubes to solve the problems.

1 The magician dropped beads into the linking circles. There are 11 beads in circle A, 11 beads in circle B, and eight beads in circle C. Five of the beads are in both circles A and B, four are in both circles B and C, and three are in both circles A and C. Two of the beads are in all three circles. How many beads did the magician drop into the circles?

2 The magician dropped coins into the linking circles. There are 11 coins in circle A, 19 coins in circle B, and 16 coins in circle C. Seven of the coins are in both circles A and B, eight are in both circles B and C, five are in both circles A and C. Three of the coins are in all three circles. How many coins did the magician drop into the circles?

Beginning Algebra Thinking, Grades 5-6 • ©Ideal School Supply Company

Linking Circles - 4

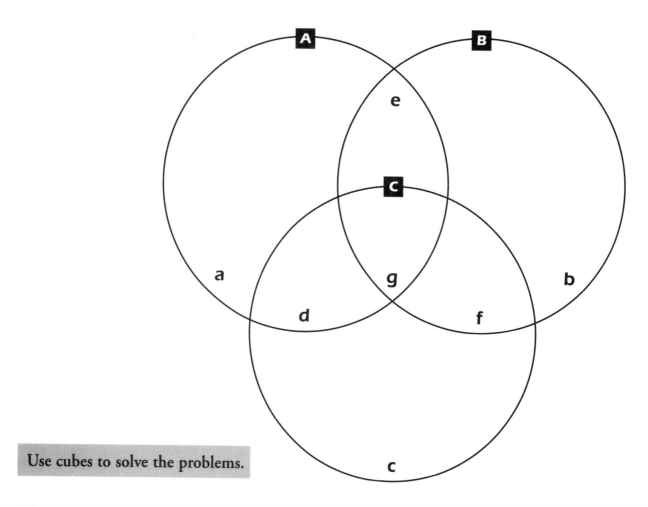

Use cubes to solve the problems.

1 The magician dropped cards into the linking circles. There are 22 cards in circle A, 20 cards in circle B, and 20 cards in circle C. Twelve of the cards are in both circles A and B, eight are in both circles B and C, and nine are in both circles A and C. Five of the cards are in all three circles. How many cards did the magician drop into the circles?

2 The magician dropped hats into the linking circles. There are 27 hats in circle A, 27 hats in circle B, and 25 hats in circle C. Nine of the hats are in both circles A and B, 10 are in both circles B and C, and 12 are in both circles A and C. Four of the hats are in all three circles. How many hats did the magician drop into the circles?

Linking Circles - 5

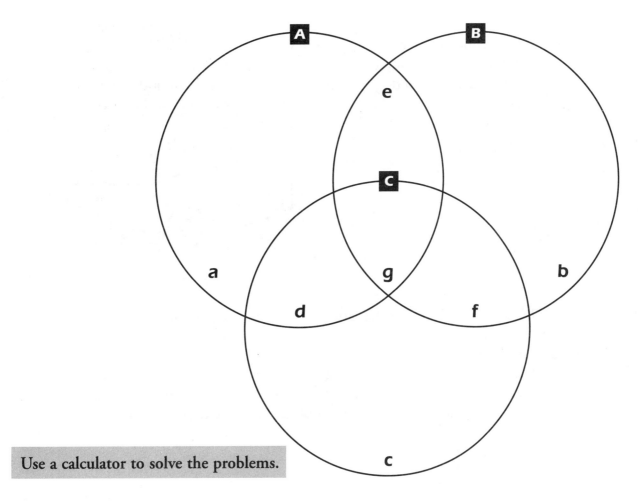

Use a calculator to solve the problems.

1 The magician dropped charms into the linking circles. There are 50 charms in circle A, 67 charms in circle B, and 70 charms in circle C. Twenty-four of the charms are in both circles A and B, 30 are in both circles B and C, and 22 are in both circles A and C. Twelve of the charms are in all three circles. How many charms did the magician drop into the circles?

2 The magician dropped coins into the linking circles. There are 186 coins in circle A, 215 coins in circle B, and 213 coins in circle C. Fifty-seven of the coins are in circles A and B, 73 coins are in circles B and C, and 80 coins are in circles A and C. Twenty-five of the coins are in all three circles. How many coins did the magician drop into the circles?

Beginning Algebra Thinking, Grades 5-6 • ©Ideal School Supply Company

Linking Circles - 6

1 Marco and Sally were doing a survey of pickup trucks. They stood at the corner of 6th and Main Streets and counted pickup trucks with camper shells, CB antennas, and dogs in the truck. They counted 17 trucks with camper shells, 12 trucks with CB antennas, and six trucks with dogs. There were seven trucks with both campers and CBs, and four trucks with both CBs and dogs. How many trucks did they count all together?

2 "Oh no!" yelled Sophia, "Who did the laundry?" No one answered, but they all rushed to look at their socks which were once white. Now some were blue, some were red, and some were yellow. There were 23 socks that were blue, 27 socks that were red, and 22 socks that were yellow. There were nine socks that were both blue and red, 10 socks that were both yellow and red, and seven socks that were both blue and yellow, and four socks that were all three colors. How many pairs of socks were in the wash?

3 It was a slow night at Bargain Movies. There were 27 people who saw "Home with the Dog," 25 people who watched "Robocat," and 34 people who saw "March of the Giant Spiders." There were eight people who saw both "Home With the Dog" and "Robocat," 13 people who saw "Robocat" and "March of the Giant Spiders," and 12 people who saw "March of the Giant Spiders" and "Home With the Dog." Then there were five people who sat through all three movies! How many people went to the movies this evening?

What's the Rule? - 1

Each machine has a secret rule.
After each problem, the rules change.
Use cubes to help find the secret rules.

1

In ⊞	4	8	12	15	20	26	30
Out ⊞	6	10					
In ▨	6	10					
Out ▨	5	9	13				

⊞ Rule: _____

▨ Rule: _____

2

In ⊞	2	4	7	10	15	24	28
Out ⊞							
In ▨	1						
Out ▨	6	8	11				

⊞ Rule: _____

▨ Rule: _____

Beginning Algebra Thinking, Grades 5-6 • ©Ideal School Supply Company

What's the Rule? - 2

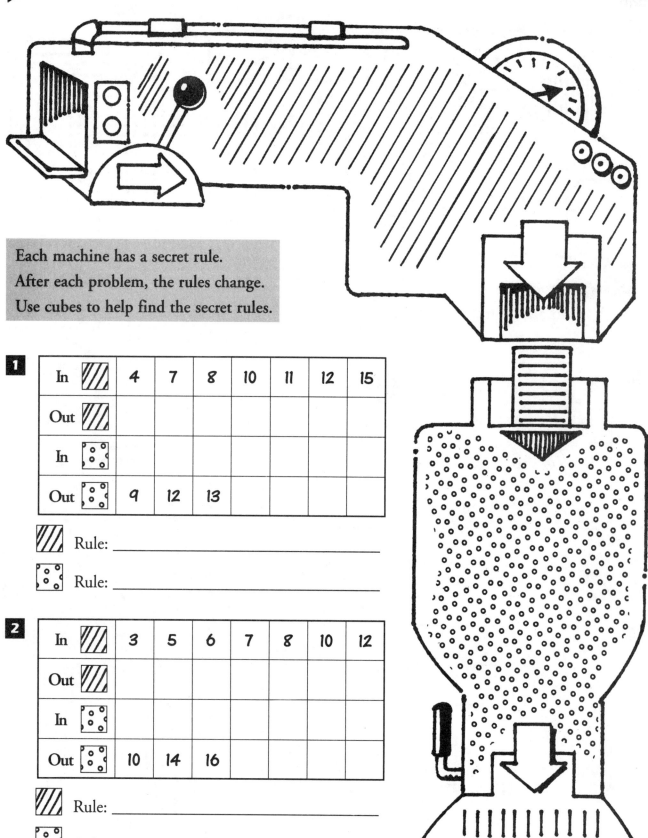

Each machine has a secret rule.
After each problem, the rules change.
Use cubes to help find the secret rules.

1

In ▨	4	7	8	10	11	12	15
Out ▨							
In ⚃							
Out ⚃	9	12	13				

▨ Rule: _____

⚃ Rule: _____

2

In ▨	3	5	6	7	8	10	12
Out ▨							
In ⚃							
Out ⚃	10	14	16				

▨ Rule: _____

⚃ Rule: _____

44 Beginning Algebra Thinking, Grades 5-6 • ©Ideal School Supply Company

What's the Rule? - 3

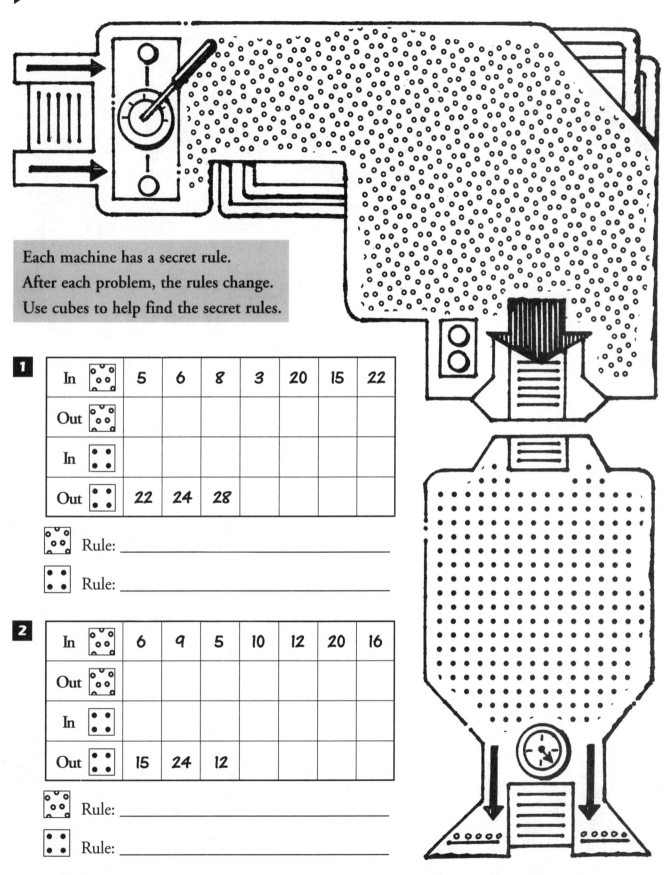

Each machine has a secret rule.
After each problem, the rules change.
Use cubes to help find the secret rules.

1

In ⚄	5	6	8	3	20	15	22
Out ⚄							
In ⚃							
Out ⚃	22	24	28				

⚄ Rule: _____

⚃ Rule: _____

2

In ⚄	6	9	5	10	12	20	16
Out ⚄							
In ⚃							
Out ⚃	15	24	12				

⚄ Rule: _____

⚃ Rule: _____

Beginning Algebra Thinking, Grades 5-6 • ©Ideal School Supply Company

What's the Rule? - 4

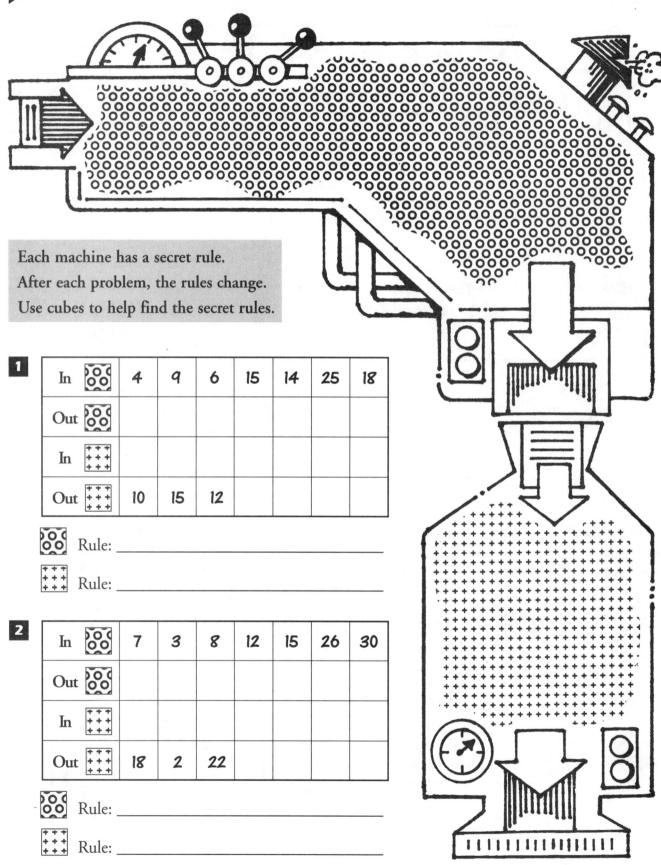

Each machine has a secret rule.
After each problem, the rules change.
Use cubes to help find the secret rules.

1

In ⬚	4	9	6	15	14	25	18
Out ⬚							
In ✚							
Out ✚	10	15	12				

⬚ Rule: _____
✚ Rule: _____

2

In ⬚	7	3	8	12	15	26	30
Out ⬚							
In ✚							
Out ✚	18	2	22				

⬚ Rule: _____
✚ Rule: _____

What's the Rule? - 5

Each machine has a secret rule.
After each problem, the rules change.
Use a calculator to help find the secret rules.

1

In ⚃	3	6	10	22	8	12	20
Out ⚃							
In ⚄							
Out ⚄	40	55	75				

⚃ Rule: _____

⚄ Rule: _____

2

In ⚃	4	7	15	6	18	26	30
Out ⚃							
In ⚄							
Out ⚄	124	214	454				

⚃ Rule: _____

⚄ Rule: _____

Beginning Algebra Thinking, Grades 5-6 • ©Ideal School Supply Company

What's the Rule? - 6

1 Every box of Fuzzy Wudgets goes to stations X and Y before it gets to the Shipping Department at the factory. A box of five Fuzzy Wudgets arrives at station X. The box contains 20 Wudgets after it leaves station Y. Then a box of 10 Fuzzy Wudgets arrives at station X. The box contains 30 Wudgets after it leaves station Y. Next, a box of 20 Wudgets arrives at station X. The box contains 50 Wudgets after it leaves station Y. What is happening at stations X and Y? If a box of 35 Wudgets goes through stations X and Y, how many Wudgets will be in the box when it gets to Shipping?

2 The Bing Bank's motto is "We do more for you." Today, the bank seems to be living up to its motto. First in line to deposit money at the automatic teller was Mary Lou. She deposited $10 and the bank slip said she deposited $34. Next in line was Raul. He deposited $23 and the bank slip said he deposited $73. Then it was Jordan's turn. He deposited $28 and the bank slip said he deposited $88. What is going on? If someone deposited $85, what would the bank slip say?

3 "Guess the Goof's puzzle!" It's that time on Teddy's Show. Someone puts balls into the Goof, then the Goof sends back balls. If you guess what rule the Goof is using, you get a prize. Today, the Goof gets eight balls and sends back six balls. Next the Goof gets 12 balls and sends back 18 balls. Then the Goof gets 15 balls and sends back 27 balls. What would happen if the Goof got 25 balls?

Story Problems - 1

1. Anita is stocking the shelves at the Stop for Pop Shop. Anita puts out twice as many bottles of lemon as lime, 10 more bottles of lime than grape, four times as many bottles of grape as raspberry, two fewer bottles of raspberry than strawberry, and nine bottles of strawberry. How many bottles of soda did Anita put on the shelves?

2. Misha has a hidden number puzzle for Cyrena. On the green paper are two triangles and one square. On the yellow paper is one triangle and one circle. The same shapes hide the same numbers. Different shapes hide different numbers. The sum of the numbers on the green sheet is the same as the sum of the hidden numbers on the yellow sheet. The sum of all the numbers is 24. What are the hidden numbers? What are the possible answers?

3. Ariana was trying to win a giant panda at the Bear Toss booth. She has to hit 10 small bears and 12 large bears to win the panda. Every large bear is worth 15 points and each small bear is worth 25 points. Ariana hit 22 bears all together, for a total of 410 points. Did she win the panda?

Story Problems - 2

1 Della and Jane are making bows for the school fair decorations. In the first hour Della makes 15 bows and Jane makes 10 bows. In the second hour Della makes 22 bows and Jane makes 19 bows. In the third hour Della makes 29 bows and Jane makes 28 bows. In the fourth hour Della makes 36 bows and Jane makes 37 bows. If these patterns continue, how long will it take them to make 285 bows?

2 Jim is working at the golf range. He is collecting yellow and white balls. There are 10 yellow balls in the can when he starts. He collects more white and yellow balls until he has 75 balls all together. If there are at least four times as many white balls as yellow balls, how many balls of each color could there be?

3 The big party at Lily Pond was in honor of Fernando Frog. The neighborhood frogs, toads, fish, raccoons, and beavers had been invited. One-fourth of the guests were toads, three-eighths were frogs, three-sixteenths were fish, and one-thirty-second were beavers. There were three beavers at the party. How many raccoons were there? How many guests were at the party?

Story Problems - 3

1 Meg and Ira bought several boxes of treasure at the garage sale. When they got home, they sorted their treasures. There were six times as many trading cards as comics, three times as many comics as books, three more books than games, two fewer games than stuffed animals, and six stuffed animals. How many treasures did they get all together?

2 Ida's Incredible Ice Cream was giving away samples of new flavors: mint magic, chips galore, and peanut madness. Forty-five people had mint, 56 had peanut, and 63 had chips. There were 18 that sampled both mint and peanut, 26 that tasted both chips and peanut, and 20 that had both chips and mint. Eight sampled all three flavors. How many people tried the new flavors?

3 Jordan Junior High students had a Saturday car wash. They charged $3.50 for big cars and $3.00 for small cars. They had a successful day and made $152.50, washing a total of 46 cars. How many little cars and how many big cars did they wash?

Beginning Algebra Thinking, Grades 5-6 • ©Ideal School Supply Company

Story Problems - 4

1 Lucia, Andy, and Karen have 150 cupcakes to decorate for the big family reunion. In the first hour Lucia decorates 5, Andy 2, and Karen 15. In the second hour Lucia decorates 7, Andy 5, and Karen 14. In the third hour Lucia decorates 9, Andy 8, and Karen 13. In the fourth hour Lucia finishes 11, Andy 11, and Karen 12. If these patterns continue and they work for five hours, will they finish the cupcakes?

2 The students at King School were asked to vote for the animal that they would like to have as a pet. There were twice as many votes for cats as dogs, four times as many votes for dogs as fish, five more votes for fish than birds, and three more votes for birds than rabbits. There were six votes for rabbits. How many votes were there for each animal?

3 Two classes are picking colors for their Fall Festival Day. The top two colors will be used for posters and invitations. The students narrowed the choices down to red, green, gold, and purple. When they voted on these colors, one-third of the students voted for red, two-sixths for gold, and two-twelfths for green. Ten students voted for green. How many students voted for purple? What two colors were chosen?

52   Beginning Algebra Thinking, Grades 5-6 • ©Ideal School Supply Company

Story Problems - 5

1 There are 55 dragons gathering for the Big Games being held at the castle today. There are three teams of dragons competing, with at least 10 dragons on each team. The teams are wearing different colors. There are at least twice as many dragons wearing purple as red and at least three more dragons wearing red than blue. How many dragons could be on each team?

2 It was the end of the Saturday Games and Puzzles Show. Odetta announced the Puzzler. Out came someone dressed as a silly creature. The Puzzler asked the audience for a number, and wrote it on a big board. Then he wrote his number to the right of it. After three numbers, the board looked like this:

Your Number	My Number
12	14
20	18
34	25

Then Odetta asked: "Can you solve the Puzzler's Puzzle today? If your number is 46, what will the Puzzler's number be?"

3 Laura and Lupe are dividing up the bags of cookies so that they each can take home the same amount of cookies. There are a total of 36 cookies. Lupe loves chocolate, so she takes the two bags of chocolate cookies along with one bag of lemon. Laura likes lemon and cinnamon, so she takes two bags of lemon and one bag of cinnamon. Every bag of chocolate has the same number of cookies in it. Every bag of lemon has the same number of cookies in it. There is a different number of each kind of cookie. How many cookies could there be in each bag?

Beginning Algebra Thinking, Grades 5-6 • ©Ideal School Supply Company

Story Problems - 6

1 Poor Joe is having a terrible time at Heavenly Donuts today. He has a machine that shoots out fresh warm donuts for each order. But today something is wrong! Joe punched in the first order for eight donuts, and out came 18 donuts. Then Joe punched in nine and out came 24 donuts. Next Joe punched in 12 and out came 42 donuts. What is going on? If Joe punched in 16, how many donuts would come out?

2 The Martin School students raised money and went to Fantasy World. There were 51 students who rode the Giant Dipper, 50 the Waterfall, and 54 the Ghost Ride. Twenty of the students didn't go on any of these rides. There were 13 students who went on the Giant Dipper and the Waterfall, 10 who went on the Waterfall and Ghost Ride, and 15 who went on the Dipper and the Ghost Ride. There were nine students who went on all three rides! How many students spent the day at Fantasy World?

3 The library did a month-long survey to see what kinds of books people liked to read. People voted for novels, mysteries, biographies, or poetry. When the votes were counted, three-fifths were for novels, two-tenths were for mysteries, and three-twentieths were for biographies. There were 18 votes for biographies. How many people voted for poetry? How many people voted?